SPACE FRONTIERS

Living and Working in Space

Helen Whittaker

A+

Smart Apple Media
P.O. Box 3263
Mankato, MN, 56002

First published in 2010 by
MACMILLAN EDUCATION AUSTRALIA PTY LTD
15–19 Claremont Street, South Yarra 3141

Visit our website at www.macmillan.com.au or go directly to www.macmillanlibrary.com.au

Associated companies and representatives throughout the world.

Copyright © Macmillan Publishers Australia 2010

Library of Congress Cataloging-in-Publication Data

Whittaker, Helen, 1965-
    Living and working in space / Helen Whittaker.
        p. cm. — (Space frontiers)
    Includes index.
    ISBN 978-1-59920-571-7 (lib. bdg.)
    1. Astronautics—Human factors—Juvenile literature. I. Title.
    TL1500.W497 2011
629.45—dc22
                                        2009038477

Edited by Laura Jeanne Gobal
Text and cover design by Cristina Neri, Canary Graphic Design
Page layout by Cristina Neri, Canary Graphic Design
Photo research by Brendan and Debbie Gallagher
Illustrations by Alan Laver

Manufactured in China by Macmillan Production (Asia) Ltd.
Kwun Tong, Kowloon, Hong Kong
Supplier Code: CP December 2009

Acknowledgments
The author and the publisher are grateful to the following for permission to reproduce copyright material:

Front cover photos of Flight Engineer Sunita L. Williams on an EVA on the *International Space Station*, courtesy of NASA/HSF; blue nebula background © sololos/iStockphoto.

Photographs courtesy of:
NASA/HSF, **3** (both), **5**, **9**, **10**, **12**, **13** (both), **14**, **15**, **16**, **17**, **18**, **19** (top), **20**, **21**, **22**, **23**, **25**, **28**, **29**, **30**, back cover; ASA/JPL-Caltech/Potsdam Univ, **6–7**; NASA/JSC, Eugene Cernan, **4**; NASA/MSFC, **24**, **26**, **27**; NASA/MSFC, Terry Leibold, **11**; Photolibrary/Science Photo Library, **8**.

Images used in design and background on each page © prokhorov/iStockphoto, Soubrette/iStockphoto.

While every care has been taken to trace and acknowledge copyright, the publisher tenders their apologies for any accidental infringement where copyright has proved untraceable. Where the attempt has been unsuccessful, the publisher welcomes information that would redress the situation.

CONTENTS

Glossary Words

When a word is printed in **bold**, you can look up its meaning in the Glossary on page 31.

SPACE FRONTIERS

A frontier is an area that is only just starting to be discovered. Humans have now explored almost the entire planet, so there are very few frontiers left on Earth. However, there is another frontier for us to explore and it is bigger than we can possibly imagine—space.

Spaceflight is one way of exploring the frontier of space. Astronaut Harrison Schmitt collects Moon rocks during the Apollo 17 mission in December 1972.

Where Is Space?

Space begins where Earth's **atmosphere** ends. The atmosphere thins out gradually, so there is no clear boundary marking where space begins. However, most scientists define space as beginning at an altitude of 62 miles (100 km). Space extends to the very edge of the universe. Scientists do not know where the universe ends, so no one knows how big space is.

Exploring Space

Humans began exploring space just by looking at the night sky. The invention of the telescope in the 1600s and improvements in its design have allowed us to see more of the universe. Since the 1950s, there has been another way to explore space—spaceflight. Through spaceflight, humans have **orbited** Earth, visited the Moon, and sent space probes, or small unmanned spacecraft, to explore our **solar system**.

LIVING AND WORKING IN SPACE

Manned spaceflights are an important part of space exploration. Crews on space shuttle flights, which are operated by the National Aeronautics and Space Administration (NASA), spend about two weeks in space. Crews on the International Space Station (ISS) spend up to six months living and working in space.

Why Send People to Live and Work in Space?

Although machines are becoming more and more sophisticated, there are still many jobs in space that only a human can do. One of the purposes of sending people to live and work in space is to study the effects of long periods of spaceflight on the human body, in preparation for future manned flights to Mars and beyond.

The Challenges of Living and Working in Space

Microgravity, which astronauts experience during a spaceflight, makes many tasks more difficult than they would be on Earth. It also affects the astronauts' bodies, causing muscles to waste away and bones to lose density. Some of the other challenges of living and working in space are cramped conditions, feeling alone, and a limited supply of water and other essentials.

Astronauts carry out important research in space to help us understand more about the universe and also improve life on Earth. ▶

Animals and Plants in Space

Humans are not the only living things to experience space. Spiders, ants, bees, and fish are also sent into space so scientists can find out how microgravity affects them. Plants are grown in space, too. When humans eventually make the long journey to Mars, they will need to be able to grow food on their spacecraft.

A TIMELINE OF LIVING AND WORKING IN SPACE

The timeline below shows every space program that has sent human beings into space.

1960	1965	1970	1975	1980

Vostok 1961–63
Description: The **Soviet Union**'s first manned spaceflight program
Flights: Six of 11 flights carried humans. Four earlier flights carried animals, including dogs, mice, rats, and a guinea pig.
Achievements: Yuri Gagarin became the first man in space in April 1961 on Vostok 1. Valentina Tereshkova became the first woman in space in June 1963 on Vostok 6.

Project Mercury 1959–63
Description: The United States's first manned spaceflight program
Flights: Out of 26 flights, only the final 6 carried astronauts. Four of the test flights carried animals.
Achievements: American astronaut Alan Shepard became the second person in space in May 1961, flying *Freedom 7*.

Voskhod 1964–65
Description: A short-lived program that made important advances in manned spaceflight
Flights: There were one unmanned and two manned flights.
Achievements: Voskhod 1, launched in October 1964, was the first spaceflight to carry more than one person. It had a crew of three. Alexei Leonov made the world's first ever space walk in March 1965 on Voskhod 2.

Project Gemini 1965–66
Description: Developed technology needed to send humans to the Moon
Flights: Ten of 12 flights were manned.
Achievements: In March 1966, *Gemini 8* became the first spacecraft to **dock** with another spacecraft in space.

Apollo Program 1963–72
Description: The only program to date that has sent humans to the Moon
Flights: There were 6 unmanned test flights and 11 manned missions, of which 6 landed on the Moon.
Achievements: Apollo 11 astronauts Neil Armstrong and Edwin "Buzz" Aldrin became the first human beings to walk on the Moon in July 1969.

Soyuz 1966–present
Description: Originally intended to send humans to the Moon, but since 1971, it has transported crews to and from space stations
Flights: There have been 26 unmanned test flights and 101 manned flights to date.
Achievements: Soyuz is the longest-running manned spaceflight program.

Space Transportation System (space shuttle) 1972–2010
Description: A reusable spacecraft designed for use in low Earth orbit
Flights: There were many suborbital test flights, some manned, others not, and, to date, 132 manned orbital flights have been completed.
Achievements: The space shuttle was the first reusable spacecraft and the first to land like an airplane.

Salyut 1971–82
Description: A series of Soviet space stations
Achievements: *Salyut 1*, launched in April 1971, was the world's first space station.

Skylab 1973–79
Description: The United States's first space station
Achievements: *Skylab* was only occupied for 171 days, but visiting astronauts spent about 2,000 hours conducting medical and scientific experiments.

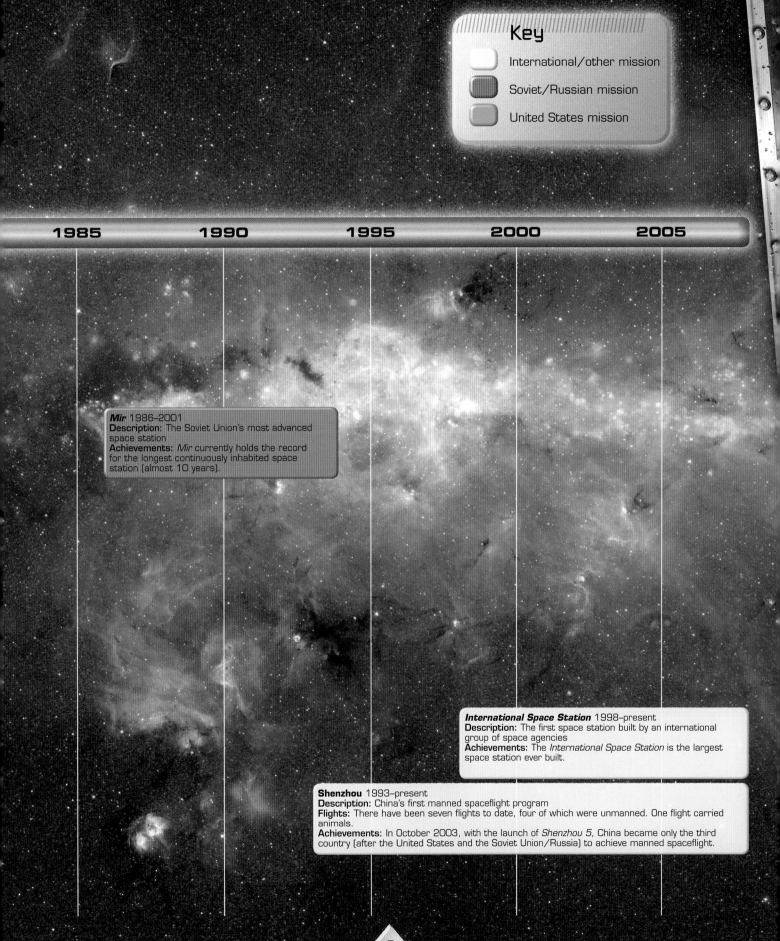

International/other mission

Soviet/Russian mission

United States mission

1985   1990   1995   2000   2005

Mir 1986–2001
Description: The Soviet Union's most advanced space station
Achievements: *Mir* currently holds the record for the longest continuously inhabited space station (almost 10 years).

International Space Station 1998–present
Description: The first space station built by an international group of space agencies
Achievements: The *International Space Station* is the largest space station ever built.

Shenzhou 1993–present
Description: China's first manned spaceflight program
Flights: There have been seven flights to date, four of which were unmanned. One flight carried animals.
Achievements: In October 2003, with the launch of *Shenzhou 5*, China became only the third country (after the United States and the Soviet Union/Russia) to achieve manned spaceflight.

Not long after the first manned spaceflight in 1961, which lasted less than two hours, humans were living in space for days and even weeks at a time. These days, crews on the International Space Station spend up to six months living and working in space.

The First Humans in Space

By 1962, astronauts from the Soviet Union were spending several days in orbit and, by 1965, American astronauts were spending up to two weeks at a time in space. One problem on these early spaceflights was space sickness.

Back then, missions had no more than three crew members, so one person falling ill for a few days would disrupt the entire mission.

Space Sickness

About half of all space travelers suffer space sickness, or Space Adaptation Syndrome, to some degree. Symptoms of space sickness include dizziness, headaches, **nausea**, and vomiting. Fortunately, the effects usually last only two or three days.

◄ Russian astronaut Valentina Tereshkova was the first woman in space. Her flight on board Vostok 6 was launched on June 16, 1963. She spent almost three days in orbit.

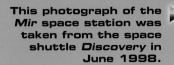

This photograph of the *Mir* space station was taken from the space shuttle *Discovery* in June 1998.

The First Space Stations

The world's first space station was the Soviet Union's *Salyut 1*, launched in 1971. The United States's first space station was *Skylab*, launched in 1973. Early space stations were made up of a single module and conditions on board were very cramped. The first multimodule space station was the Soviet Union's *Mir*, which was constructed in space between 1986 and 1996.

The International Space Station

Sixteen countries are working together to complete the *International Space Station*, which is the only space station currently in operation. Its main purpose is to carry out scientific research. The *ISS* is made up of 15 separate modules, including laboratories, docking ports, **airlocks**, and living quarters.

The table below shows how far space stations have come since the days of *Salyut 1*. As the number of modules increased, so did the living and working areas, shown below as living volume. More astronauts can be sent on missions to the *ISS* now to conduct construction work, repairs, and experiments.

Space station	Salyut 1	Skylab	Mir	ISS
Number of modules	1	1	7	15
Weight (in pounds/kilograms)	40,620 lb (18,425 kg)	199,750 lb (90,605 kg)	274,123 lb (124,340 kg)	669,462 lb (303,663 kg)
Living volume (in cubic feet/cubic meters)	3,500 ft³ (99 m³)	12,700 ft³ (360 m³)	12,360 ft³ (350 m³)	12,643 ft³ (358 m³)
Crew	3	3	3	6

TRAINING TO LIVE AND WORK IN SPACE

Astronauts train for a very long time. They must know their spacecraft completely and know what to do in an emergency. They also need to practice moving and working in a spacesuit.

Getting to Know the Spacecraft

During training, astronauts spend hundreds of hours in simulators. A simulator is a full-size model of a spacecraft. Astronauts learn how to operate all the spacecraft's systems and practice dealing with emergencies while in the simulator. Simulations can also take the form of **virtual reality** computer programs, which astronauts interact with using special headsets and gloves.

▼ **American astronaut Joseph Acaba trains with a virtual reality program at Johnson Space Center in the United States before his mission on the space shuttle *Discovery* in 2009.**

The Five Areas of Astronaut Training

Astronauts undergo training in five important areas.

- Classroom work
  To study **aerodynamics**, physics, human biology, and computer science.
- Flight training
  To learn to fly an airplane.
- Survival training
  To learn how to survive after an unplanned landing in water or on land.
- Basic mission training
  To get to know the spacecraft and prepare for living and working in **microgravity**.
- Advanced mission training
  To practice specific mission tasks and deal with emergency situations.

Astronauts train underwater at Marshall Space Flight Center in the United States to test parts for the *International Space Station*. Divers are in the water to offer help if astronauts need it.

The Vomit Comet

The Vomit Comet refers to any aircraft that is used to provide a weightless environment. It does this by following a **parabolic** flight path, which leaves its occupants in a state of **free fall** for about 25 seconds out of every 65. This process generally makes most of its occupants experience nausea.

Training for Weightlessness

To be able to live and work in space, astronauts need to do more than just learn new skills. They also need to prepare for microgravity. This means they have to relearn how to do the simple things they do every day, but while weightless. To do so, they experience weightlessness in an airplane nicknamed the Vomit Comet. They also learn to scuba dive.

Training for Space Walks

To practice moving around and working in microgravity while wearing a spacesuit, astronauts train in a large swimming pool. Their spacesuits are weighted to give them **neutral buoyancy** in the water. This simulates the feeling of weightlessness. While underwater, the astronauts practice the tasks they will perform while working outside the spacecraft.

FOOD IN SPACE

Organizing the food for any expedition is a challenge. Due to the unique conditions in space, organizing the food for a space mission is even more challenging.

Difficulties with Food in Space

Cooking food in space is almost impossible—weightless eggs and weightless bacon will not stay on the frying pan! Instead, astronauts reheat precooked food and rehydrate dried food by adding hot water. It is important to avoid foods that crumble because crumbs might float away and get stuck inside equipment. For this reason, salt, pepper, and other condiments are in liquid form.

Types of Space Food

The table below shows the main types of food eaten in space.

American astronauts Shane Kimbrough and Sandra Magnus attempt to catch their fruit on board the space shuttle *Endeavour*.

Food Category	Explanation	Examples
fresh	food that is not processed in any way	apples and bananas
frozen	quick-frozen to prevent ice crystals from forming and to preserve taste	quiches, casseroles, and chicken pot pie
intermediate moisture	partially dried food	dried peaches, dried apricots, and beef jerky
irradiated	cooked and packed in foil pouches, and **sterilized** using **ionizing radiation**	beef steak and roast turkey
natural form	untreated food sealed in foil pouches	biscuits, nuts, and muesli bars
rehydratable	dried food that can be prepared by adding hot water	oatmeal
thermostabilized	heat-treated and sealed in cans or containers	tuna, lamb with vegetables, and puddings

▲ Japanese astronaut Akihiko Hoshide prepares to have a meal on the space shuttle *Discovery*.

Food is stored in different packaging as it has to be prepared in different ways. ▶

Meals in Space

Some space foods can be eaten just as they are, but others need to be heated in a food warmer. Drinks and rehydratable food are prepared by adding water and mixing. Astronauts use a meal tray with Velcro strips to hold their food in place. After the meal, they clean their trays and eating utensils with wet wipes.

Did You Know?

On the *International Space Station*, crew members do not sit down to eat. Instead, they simply float next to the table.

Space Menus

Here is a typical dinner menu on the *International Space Station*.

Menu Item	Type of Food
prawn cocktail	rehydratable
steak	irradiated
macaroni and cheese	rehydratable
fruit cocktail	thermostabilized
strawberry drink	beverage
tea with lemon	beverage

HOUSEWORK IN SPACE

Just like people on Earth, astronauts have household chores. In fact, it is even more important that they stay on top of these chores, because their lives may depend on it.

Cleaning

The crew of the *International Space Station* follows a strict cleaning schedule to keep **microorganisms**, such as viruses, bacteria, and fungi, at bay. Microorganisms can affect the health of the crew and can "eat" their way through the hard surfaces of a spacecraft.

American astronaut Sunita Williams collects a sample of air in the Destiny laboratory of the *International Space Station*. This is part of the procedure known as SWAB (Surface, Water, and Air Biocharacterization).

SWAB (Surface, Water, and Air Biocharacterization)

SWAB is an ongoing procedure on the *International Space Station*. It examines air, surface, and water samples for microorganisms and **allergens**. In the short term, it alerts the crew members to any potential problems, so they can take action. In the long term, it will contribute to the development of healthier spacecraft environments.

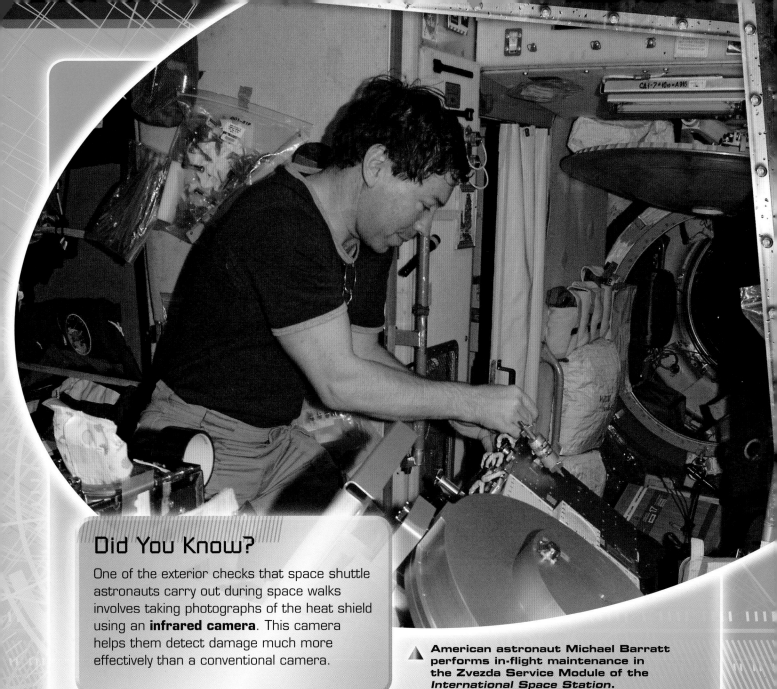

▲ American astronaut Michael Barratt performs in-flight maintenance in the Zvezda Service Module of the *International Space Station*.

Taking Out the Trash

Astronauts on short spaceflights bring all their garbage back to Earth for recycling and disposal. On the *ISS*, crew members load waste containers onto a supply vehicle docked at the station. When this vehicle is full, it undocks from the *ISS* and reenters Earth's atmosphere, burning up as it does so.

Home Maintenance

To keep the spacecraft running safely, astronauts perform daily checks of all the spacecraft's systems, including power, navigation, computers, and life-support. Then they repair or replace parts as required. They also put on their spacesuits and inspect the outside of the spacecraft to check for damage sustained during launch or caused by **micrometeoroid** impacts during orbit.

HYGIENE IN SPACE

Personal hygiene is very important in space. Astronauts work in a confined environment in close contact with each other, so infection can spread quickly. An astronaut who falls sick cannot visit a doctor or go to a hospital.

Washing

Each astronaut has his or her own personal hygiene kit. Astronauts can shave and brush their teeth normally, but there are no showers in space.

They have to take sponge baths and wash their hair with rinse-free shampoo. The bathroom on board the *International Space Station* is called the hygiene center.

▼ **American astronaut James Voss shaves with an electric razor in the Zvezda Service Module of the *International Space Station*.**

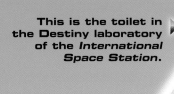

This is the toilet in the Destiny laboratory of the *International Space Station*.

Going to the Toilet

An ordinary toilet relies on **gravity** and would not work in space. Astronauts use a special space toilet. Each astronaut has his or her own personal urine funnel. When they want to sit on the toilet, they have to strap themselves down. The toilet works like a vacuum cleaner, sucking air and waste into the waste tank.

Doing the Laundry

Space is not a good place to do laundry. Space shuttle astronauts bring enough clothes with them to last the entire mission, and they take their dirty laundry back home. On the *ISS*, clean clothes are provided by a supply vehicle and dirty clothes are placed in the vehicle to be burned.

Water on the *International Space Station*

Water is in limited supply on the *ISS*. All of the station's wastewater, including the crew's urine, is sent to an on-board water treatment plant, where it is thoroughly cleaned. It is then used again.

CLOTHING IN SPACE

Astronauts wear different types of clothing at different times. During launch and **reentry** they wear a pressure suit. Once in space, they wear everyday clothes. When working outside the spacecraft, they wear a spacesuit.

Pressure Suits

Astronauts wear a pressure suit during launch and reentry to give them a better chance of surviving if they have to eject from the spacecraft. The pressure suit includes oxygen tanks, a parachute, a life raft, a radio, **flares**, and enough drinking water for a day. The entire suit weighs about 80 pounds (35 kg).

▼ **American astronaut Charles Hobaugh is helped into a pressure suit during a water survival training session at Johnson Space Center.**

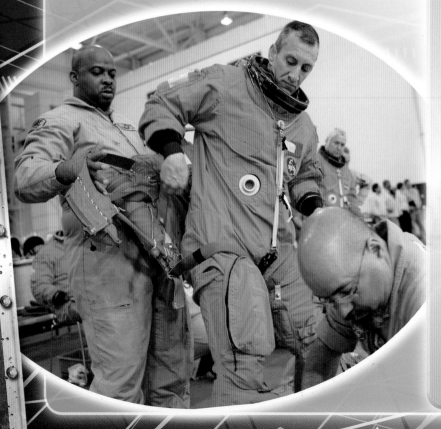

Putting on a Pressure Suit

Space shuttle astronauts spend up to 45 minutes getting into their pressure suits and testing the different parts to ensure they work.

STEP 1: In private, each astronaut puts on a thermal undersuit.

STEP 2: Technicians help the astronaut put on the pressure suit. Legs go in first, followed by arms.

STEP 3: Next, the astronaut pushes his or her head through the metal neck ring. It is a bit of a squeeze getting the astronaut's head into the suit, because there is a tight seal around the neck.

STEP 4: Technicians zip up the suit and help the astronaut into the special boots.

STEP 5: The helmet and gloves are put on and locked in place with metal rings. The suit is checked for airtightness, then the helmet and gloves are removed until the astronaut is seated in the spacecraft.

STEP 6: Each astronaut's pockets are packed with survival items such as flares and a radio.

STEP 7: The crew then travels by van to the launch pad. In the van, they can plug into a cooling unit to stop themselves from overheating in the heavy suits.

STEP 8: At the launch pad, astronauts put on their parachutes and communications headsets.

STEP 9: While the astronauts are being strapped into their seats on the spacecraft, technicians help them put on their gloves and helmets to prepare for launch.

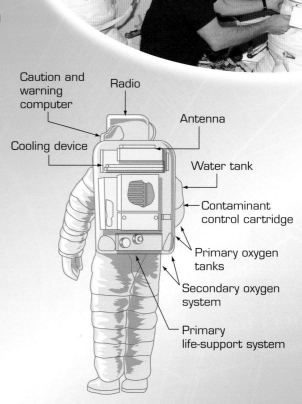

American astronauts Steven Swanson (right) and Richard Arnold (left) are helped into their spacesuits, or Extravehicular Mobility Units (EMUs).

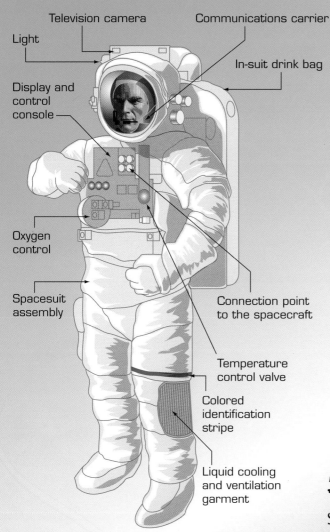

Light
Television camera
Communications carrier
In-suit drink bag
Display and control console
Oxygen control
Spacesuit assembly
Connection point to the spacecraft
Temperature control valve
Colored identification stripe
Liquid cooling and ventilation garment

Caution and warning computer
Radio
Antenna
Cooling device
Water tank
Contaminant control cartridge
Primary oxygen tanks
Secondary oxygen system
Primary life-support system

▲ This illustration shows the EMU used by NASA astronauts during space walks.

Everyday Clothes

Most of the time, astronauts wear everyday clothes such as T-shirts, shorts, pants, and sweaters. On the *International Space Station*, astronauts are not allowed to change their clothes as often as they would on Earth. They change their underwear and socks every other day and have fresh exercise clothes every three days. Their work clothes have to last for 10 days.

Spacesuits

Spacesuits are worn whenever an astronaut works outside the spacecraft. Without a spacesuit, astronauts would be unable to breathe in the vacuum of space and their bodies would start reacting to the lack of pressure outside Earth's atmosphere.

The spacesuit includes underwear with a network of tubes attached. Water flows through the tubes to keep the astronaut's body cool. The main suit, with pants, boots, upper body, sleeves, gloves, and helmet, protects the astronaut from the environment. A backpack houses the life-support system, which can be managed by a control console on the front of the suit.

SLEEPING IN SPACE

Astronauts need sleep just like anyone else, but without the force of gravity, lying down to sleep is difficult! The way astronauts sleep in space is very different from the way they sleep on Earth.

Challenges of Sleeping in Space

In orbit, sleeping astronauts would float around and bump into things, unless they were strapped in. However, even when strapped in, astronauts do not feel like they are lying down, so some of them have trouble falling asleep.

▼ Italian astronaut Paolo Nespoli (right) and American crew members Pamela Melroy (left) and George Zamka (center) take a nap on the space shuttle *Discovery*.

Sleeping on the Space Shuttle

With up to seven crew members on the space shuttle at any time and not much room to spare, sleeping conditions can be cramped. Astronauts can sleep in their seats or in a sleeping bag strapped to the wall. They are awakened every morning by a different tune broadcast by Mission Control in Texas.

Sleeping on the International Space Station

Each crew member on the *ISS* has his or her own cabin, which is about the size of a telephone booth. The cabin has a laptop computer, storage for personal belongings, and a sleeping bag strapped to the wall. Some cabins have a window with a blind. In the morning, crew members are awakened by an alarm clock.

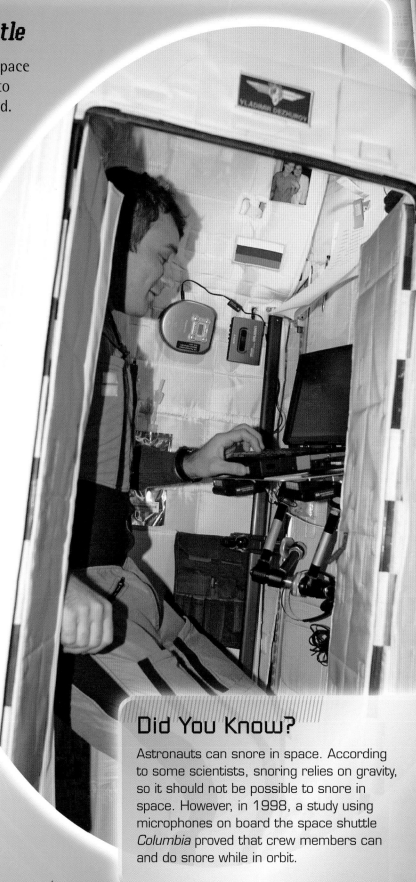

Russian astronaut Vladimir Dezhurov works on a laptop computer in his cabin on the *International Space Station*. The laptop and other equipment are attached to the wall to prevent them from floating away.

Did You Know?

Astronauts can snore in space. According to some scientists, snoring relies on gravity, so it should not be possible to snore in space. However, in 1998, a study using microphones on board the space shuttle *Columbia* proved that crew members can and do snore while in orbit.

STAYING HEALTHY IN SPACE

The human body evolved over time to cope with Earth's gravity. Weightlessness can result in many health problems connected to blood, muscles, and bones. It is, therefore, very important that astronauts look after themselves in space.

Healthy Food

An important part of staying healthy is eating the right food. Space foods are designed with nutrition in mind and are analyzed to find out their nutritional content. Menus are prepared in advance to make sure meals provide astronauts with the right balance of carbohydrates, fats, proteins, fiber, vitamins, and minerals.

Food Safety

Healthy food is not only nutritious but is safe to eat. Space food is prepared in a strictly controlled and hygienic environment to prevent the growth of bacteria or mold. If astronauts were struck by food poisoning on a spaceflight, the entire mission could be affected.

▼ **Japanese astronaut Takao Doi (seated, right) and his American counterparts Dominic Gorie (left) and Gregory H. Johnson meet with dietician Michelle Pittman to try the food and finalise the menu for their spaceflight.**

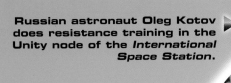
Russian astronaut Oleg Kotov
does resistance training in the
Unity node of the *International
Space Station.*

Keeping Fit

In space, muscles begin to waste away and bones become weaker because the force of gravity is not acting on the body. One way of slowing down this process is to get plenty of exercise. Astronauts spend at least two hours a day working out on specially adapted exercise equipment, such as exercise bikes and treadmills.

Avoiding Infection

Immediately before a spaceflight, astronauts spend several days in **quarantine** so that they are less likely to come into contact with bacteria or viruses that could make them fall sick during the mission. During the mission, astronauts regularly wipe surfaces with a special cleaner to prevent the growth of microorganisms.

Did You Know?

To help astronauts avoid infection, the air and water on board the *International Space Station* are purified to a very high standard. The air is a lot cleaner than the air inside a house on Earth and the drinking water is much purer than the water from a kitchen faucet.

FREE TIME IN SPACE

Astronauts work hard, but they don't work all the time. Like everyone else, they need time to relax and have fun. Mission planners make sure every astronaut gets some free time in the daily schedule.

Relaxing

One popular pastime in space is looking out the window. This is not surprising, as Earth looks beautiful from orbit and the view is constantly changing.

Astronauts can also relax by doing some of the things they might do back on Earth, such as reading a book, watching a DVD, or talking with the other astronauts.

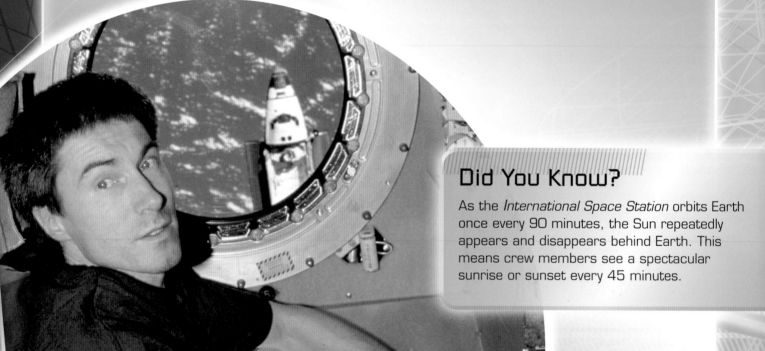

Did You Know?

As the *International Space Station* orbits Earth once every 90 minutes, the Sun repeatedly appears and disappears behind Earth. This means crew members see a spectacular sunrise or sunset every 45 minutes.

◀ **Russian astronaut Sergei Krikalev enjoys the view from a porthole on board the Zvezda Service Module of the *International Space Station*, as the space shuttle *Atlantis* approaches.**

Fun and Games

Crew members can also have fun by playing board games, card games, and ball games. A popular game on the *International Space Station* is racing from one end of the station to the other. Astronauts race against the clock rather than against each other, because the openings between the modules are so narrow.

Keeping in Touch

During their free time, astronauts can contact friends and family back on Earth via e-mail, radio transmissions, and recorded video messages. Keeping in touch is particularly important for the crew of the *ISS*, who may be away from home for up to six months at a time.

Amateur Radio

There is an amateur radio station on board the *International Space Station*, which can communicate with other amateur radio stations on the ground. Astronauts use amateur radio to answer questions from the general public, including students, and to keep in touch with friends and family.

SCIENTIFIC RESEARCH IN SPACE

On the International Space Station astronauts have a day job that involves carrying out scientific research through experiments. These experiments are one of the main reasons why the space station was built.

Why Do Research in Space?

Research helps scientists prepare for future space missions. Astronauts test the effects of long-term weightlessness on their bodies and develop ways of growing food in space to prepare for missions to Mars. New and improved materials can also be better developed in space due to the effects of microgravity.

Developing New and Improved Materials

Materials manufactured in microgravity have different physical properties than the same materials manufactured on Earth. These differences can improve the material in some way. The hormone insulin, for example, which is used to treat diabetes, is of a better quality when produced in space. In the future, huge orbiting factories and laboratories may manufacture many of the materials we need.

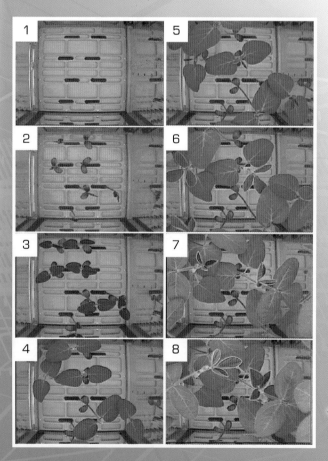

Did You Know?

Special lights called Light Emitting Diodes (LEDs) are used to grow plants in space. They provide enough light for plants to grow well, but use only a small amount of electricity. They save energy by emitting light only at the particular frequencies plants need in order to grow.

◀ **This image shows the growth of soybeans over time in the Advanced Astroculture Plant Growth Chamber of the *International Space Station*.**

Laboratories on the International Space Station

The *ISS* has four laboratory modules built by the European Space Agency and the space agencies of the United States, Russia, and Japan. Each is fully equipped with everything found in a regular laboratory, including incubators, refrigerators, freezers, gas cabinets, glove boxes, and microscopes.

Some experiments do not take place in the laboratories but outside, in the vacuum of space.

The table below shows some of the types of experiments carried out on the *ISS*.

Type of Research	What It Does	How It Is or Will Be Used
biological research	finds out how plants and animals react to microgravity and the environment of space	to ensure there is enough food on longer missions in the future
Earth observation research	monitors Earth from space	to collect atmospheric and climate data for agricultural and environmental research
human research	finds out how physical and mental health are affected by spaceflight	to ensure astronauts on longer missions in the future stay healthy
physical science research	explores the physics of microgravity	to develop new materials for use in space and on Earth
radiation measurement	finds out how much radiation the exterior of the *ISS* receives and how much reaches the crew inside	to calculate possible doses of radiation on a flight to Mars and to design future spacecraft that limit exposure to radiation

WORKING OUTSIDE
THE SPACECRAFT

During a spaceflight, not every job can be done from inside the spacecraft. On most missions one or more astronauts perform at least one space walk, or Extravehicular Activity (EVA), during which they work outside the spacecraft.

Why Work Outside the Spacecraft?

Working in a spacesuit in the vacuum of space is both difficult and dangerous. However, some tasks can only be performed by astronauts working outside the spacecraft. Such tasks include maintaining and repairing the exterior of the spacecraft, monitoring external experiments, and repairing space hardware such as **satellites** and telescopes.

Did You Know?

The longest space walk took place during mission STS-102 in 2001. American astronauts James Voss and Susan Helms spent 8 hours and 56 minutes preparing the *International Space Station* to receive a new cargo module.

▼ **American astronaut Stephen Bowen works on parts of the** *International Space Station.*

American astronaut Sunita Williams uses the Pistol Grip Tool while working on the outside of the *International Space Station* in January 2007.

The Pistol Grip Tool

Designing the multipurpose Pistol Grip Tool presented many challenges. Not only did it have to do the job of several different power tools, but astronauts had to be able to use it while wearing thick spacesuit gloves. The tool's battery and computer also had to work at extreme temperatures.

A Personal Spacecraft

Without a spacesuit, an astronaut would die after only a few minutes of exposure to the hostile environment of space. A spacesuit acts like a mini-spacecraft for one person. It protects the astronaut from extremes of temperature, radiation, bright sunlight, and bullet-like micrometeoroids. It also provides air to breathe and water to drink.

Tools in Space

One of the most useful space tools is the Pistol Grip Tool, which is an electric drill, screwdriver, and wrench. Astronauts also carry a trace gas analyzer, which detects any leaking gas or liquid such as oxygen, water, or rocket fuel. A robotic crane maneuvers large objects. It can also pick up astronauts and place them in the right position.

THE FUTURE OF LIVING AND WORKING IN SPACE

The most time a single person has spent living and working in space is 437.7 days, which is just over 14 months. In order to make a trip to Mars and back, astronauts will have to spend several years at a time in space. Farther into the future, some people may even spend their entire lives in space.

Coming Soon

More orbiting space stations are likely to be built once the *International Space Station* shuts down and, possibly, while it is still operational. NASA is planning to send humans back to the Moon by 2020 and will eventually build a permanently inhabited lunar base, where astronauts and scientists can live and work for many months or even years.

Looking Farther Ahead

The astronauts who make the first manned flights to Mars will live and work in space for several years, first on board the spacecraft and then on a base that they build on Mars. Eventually, humans may live on other planets and moons in the solar system. Children born there might spend their entire lives in space.

▼ This artist's impression shows what a lunar base might look like. In the background are the living quarters, on the left is a rover, and on the right is a landing craft.

GLOSSARY

aerodynamics
a branch of science that examines the way air interacts with moving objects, such as aircraft

airlocks
chambers in a spacecraft that allow astronauts to move in and out of the spacecraft without affecting its air pressure

allergens
substances that can cause an allergy, or bad reaction, in some people

atmosphere
the layer of gases surrounding a planet, moon, or star

dock
join one spacecraft to another

flares
devices that produce a bright light or colored smoke, which can be used to attract attention in an emergency

free fall
an object is in free fall when it is affected by only gravity and no other force

gravity
the strong force that pulls one object toward another

infrared camera
a camera that uses infrared radiation rather than visible light to create an image

ionizing radiation
radiation caused by high-energy electromagnetic waves with very short wavelengths, which can be used to sterilize food

microgravity
weightlessness, a phenomenon experienced when orbiting a planet

micrometeoroid
a tiny, fast-moving particle of dust or rock commonly found near Earth that can cause damage to spacecraft and spacesuits

microorganisms
very tiny living things, such as viruses, bacteria, and fungi, which are too small to be seen with the naked eye

nausea
the feeling that you are about to vomit

neutral buoyancy
a condition that occurs when the weight of an object in water is equal to the weight of the water it displaces, causing the object to neither sink nor rise

orbited
followed a curved path around a more massive object while held in place by gravity; the path taken by the orbiting object is its orbit

parabolic
describing a type of curve known as a parabola, which is similar to the path taken by an object that is thrown into the air and falls to the ground in a different place

quarantine
a period of time when a person is kept away from other people in order to prevent them from catching or spreading disease

reentry
the process of reentering Earth's atmosphere after a spaceflight

satellites
natural or artificial objects in orbit around another body

solar system
the Sun and everything in orbit around it, including the planets

Soviet Union
a nation that existed from 1922 to 1991, made up of Russia and 14 neighboring states

sterilized
cleaned and germ-free

virtual reality
technology that allows a user to interact with a three-dimensional environment simulated by a computer

INDEX